ENGINEERING FOR DISASTER

ENGINEERING FOR FLOODS

by Samantha S. Bell

FOCUS READERS
NAVIGATOR

WWW.FOCUSREADERS.COM

Copyright © 2021 by Focus Readers®, Lake Elmo, MN 55042. All rights reserved. No part of this book may be reproduced or utilized in any form or by any means without written permission from the publisher.

Focus Readers is distributed by North Star Editions:
sales@northstareditions.com | 888-417-0195

Produced for Focus Readers by Red Line Editorial.

Content Consultant: Faisal Hossain, Professor of Civil & Environmental Engineering, University of Washington

Photographs ©: Shutterstock Images, cover, 1, 11, 15, 17, 25, 26–27; Charlie Neibergall/AP Images, 4–5; Red Line Editorial, 7, 29; Ival Lawhon/St. Joseph News–Press/AP Images, 8–9; Laurence Fordyce/UIG/Science Source, 12–13; Wayne Parry/AP Images, 18–19; iStockphoto, 21, 23

Library of Congress Cataloging-in-Publication Data
Names: Bell, Samantha, author.
Title: Engineering for floods / by Samantha S. Bell.
Description: Lake Elmo, MN : Focus Readers, [2021] | Series: Engineering for disaster | Includes index. | Audience: Grades 4-6.
Identifiers: LCCN 2020003412 (print) | LCCN 2020003413 (ebook) | ISBN 9781644933794 (hardcover) | ISBN 9781644934555 (paperback) | ISBN 9781644936078 (pdf) | ISBN 9781644935316 (ebook)
Subjects: LCSH: Flood control--Juvenile literature.
Classification: LCC TC530 .B46 2021 (print) | LCC TC530 (ebook) | DDC 627/.4--dc23
LC record available at https://lccn.loc.gov/2020003412
LC ebook record available at https://lccn.loc.gov/2020003413

Printed in the United States of America
Mankato, MN
082020

ABOUT THE AUTHOR
Samantha S. Bell lives with her family and lots of pets in the foothills of the Blue Ridge Mountains. She is the author of more than 100 nonfiction books for students.

TABLE OF CONTENTS

CHAPTER 1
The Great Flood 5

CHAPTER 2
The Destructive Force of Floods 9

CHAPTER 3
Hard Engineering 13

CHAPTER 4
Soft Engineering 19

CASE STUDY
Hesketh Out Marsh 24

CHAPTER 5
Preparing for Future Floods 27

Focus on Engineering for Floods • 30
Glossary • 31
To Learn More • 32
Index • 32

CHAPTER 1

THE GREAT FLOOD

The Mississippi and Missouri Rivers have a history of flooding. Engineers built **levees** and dams to help control the flooding. Then, in 1993, strange weather brought heavy rain to the north-central United States. Usually, spring and summer storms last just a week or two. But this time, the storms kept coming.

Flooding along the Mississippi and Missouri Rivers affects the people living in nearby areas.

Some places received more than 4 feet (1.2 m) of rain.

As a result, more than 1,000 levees failed to hold back water or broke. The water flooded nine states for months. It covered more than 400,000 square miles (1,036,000 sq km). Many people had to leave their homes. Some could never return. And more than 50 people died.

The flood of 1993 became known as the Great Flood. It was one of the worst natural disasters in US history. The flood covered towns and ruined farmland. It destroyed water and sewage treatment plants. It closed down transportation. In total, it caused approximately $15 billion in damage. Since then, engineers have rebuilt most of the damaged levees. They also added new, higher levees and floodwalls in some areas. They wanted cities to be safe from future floods.

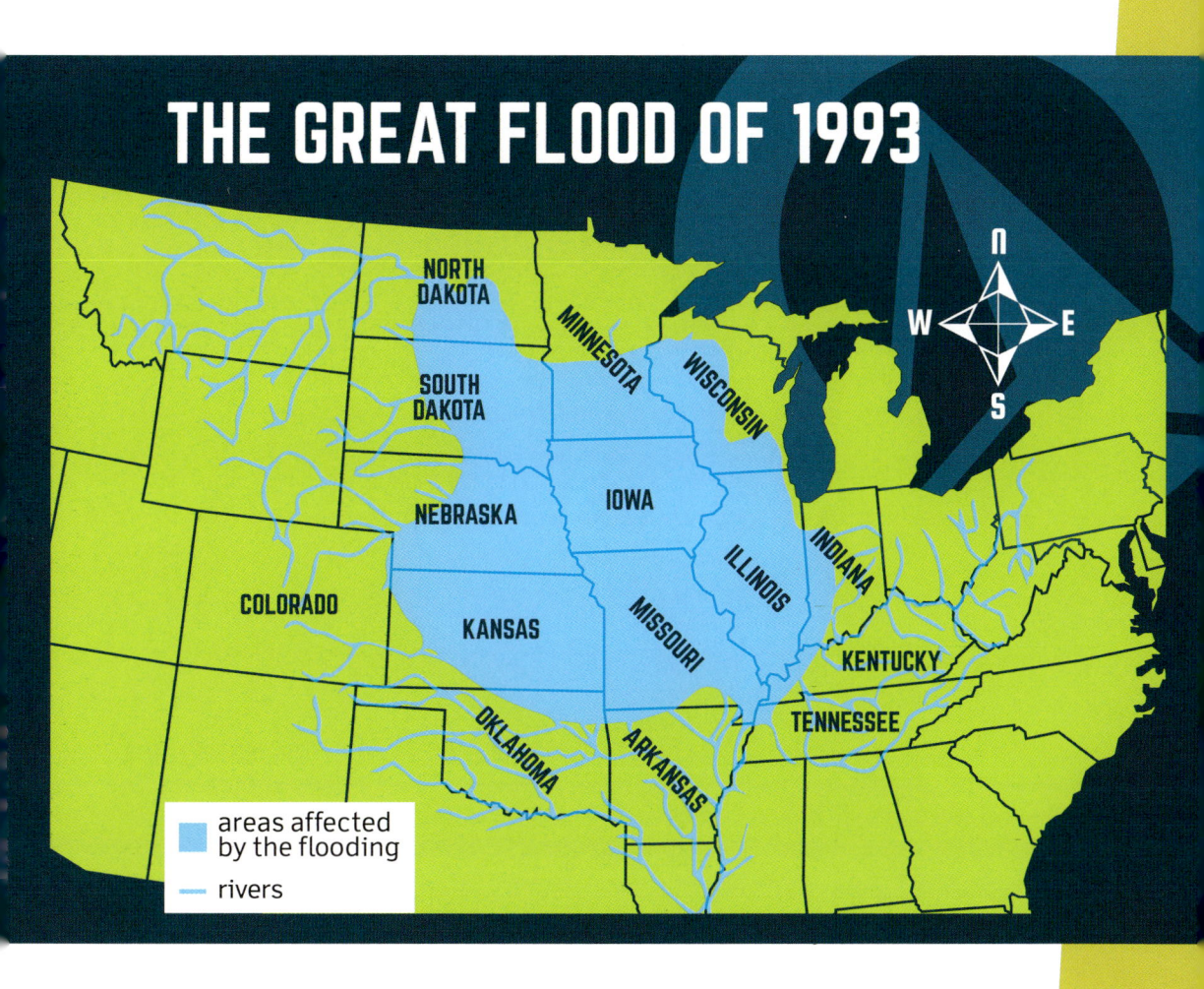

THE GREAT FLOOD OF 1993

- areas affected by the flooding
- rivers

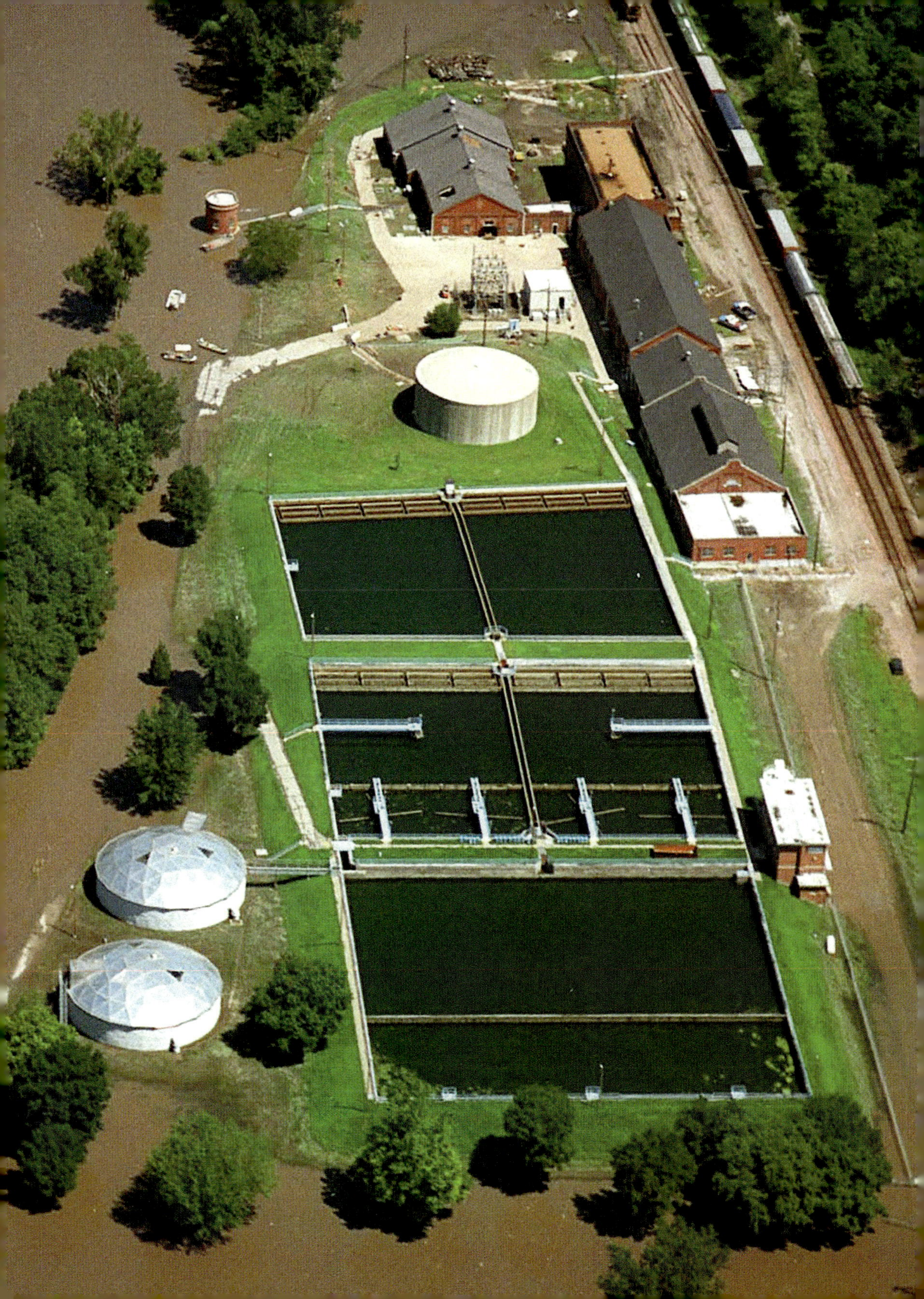

CHAPTER 2

THE DESTRUCTIVE FORCE OF FLOODS

Floods happen for many reasons. Rivers can overflow because of heavy rain. The soil around a river can only hold so much water. If the soil is already very wet, the extra water becomes runoff. The runoff quickly flows downstream and floods those areas. Snow or ice that melts quickly also can cause flooding.

The Great Flood affected water treatment plants, making tap water unsafe to drink.

Breaks in dams and barriers can cause flooding, too. And hurricanes can flood areas along the coast.

Engineers study why floods happen. They use this information to try to prevent future floods. They also try to reduce the effects of floods. For example,

IMMEDIATE DANGER

Flash floods are floods that develop within a few hours. Often, people are not prepared for these floods. They do not have time to protect their property. They can become trapped in their cars or homes. Engineers try to predict flash floods. They measure water levels in the soil and rivers. They study the land, **vegetation**, and rainfall. This way, they can warn people who are in danger.

Heavy rains can flood flat land. The ground cannot absorb the rainwater fast enough to prevent flooding.

they build structures to hold the water back. They use the land's natural features to control flooding. These efforts help keep communities safe from floods.

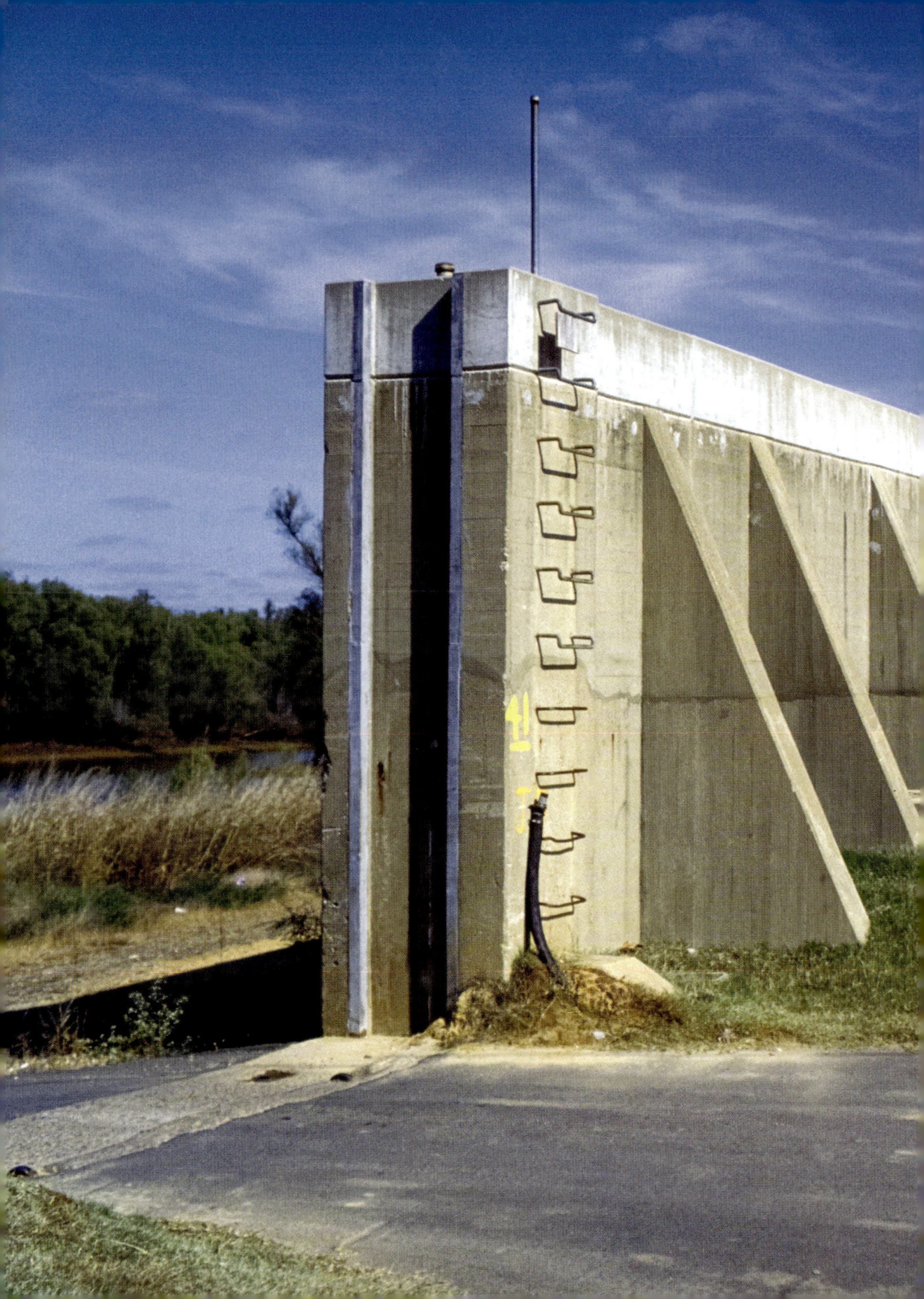

CHAPTER 3

HARD ENGINEERING

Engineers can create human-made structures to help prevent floods. They design and build the structures. They also help maintain them. These structures are known as hard engineering.

Some structures hold back floodwater. Levees are made of sand, rocks, or soil. Engineers might add concrete floodwalls.

A floodwall holds back water during floods. A gate can be opened when the water is low.

Or they might build floodgates to control the flow of water. Some of these barriers are fixed in place. Others can be moved to where they are needed most.

Other forms of hard engineering carry water away. **Channels** move floodwater back into a river. Sewer and drainage

NEW PAVEMENTS

On unpaved land, the soil can take in extra water. But on paved land, the water does not have anywhere to go. So, the water rises and floods. To fix this problem, engineers created a new type of pavement. The pavement has many tiny holes. Rainwater can flow through the holes into the ground below. The soil can then absorb the water slowly.

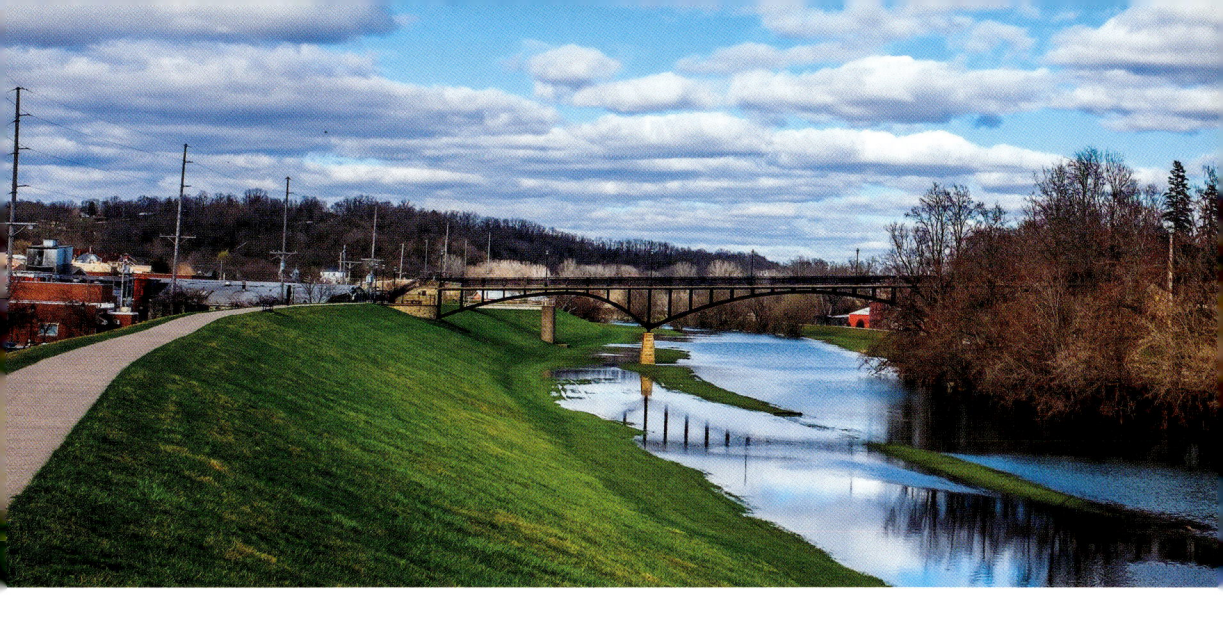

People can walk along the top of this levee. It protects the city of Galena, Illinois.

pipes carry water out of a city. For example, in Tokyo, Japan, engineers created a series of tunnels. The tunnels carry floodwater to a large tank. Pumps send the floodwater into the Edo River system. The water flows into Tokyo Bay.

Sometimes, engineers build a dam to block a river. This action creates a human-made lake behind the dam.

During heavy rains, rainfall collects in the lake instead of flooding. Later, engineers can release the extra water slowly.

Engineers also can change the river itself. They might dredge it by removing sand from the bottom. This action makes the river deeper. The river can then carry more water downstream. Engineers also might straighten a river. They cut channels to connect parts of the river. Then the water flows faster. It can pass through an area more quickly.

Hard engineering does not always fix flooding problems. Sometimes, the barriers break. Drainpipes can become clogged with leaves and other objects.

The Hoover Dam was built in the 1930s to control the flooding of the Colorado River.

Then floodwater cannot flow through them. Dredging has to be done again and again. Changing the flow of a river can destroy the homes of animals and plants. It also can cause more flooding downstream. For these reasons, engineers often use other methods.

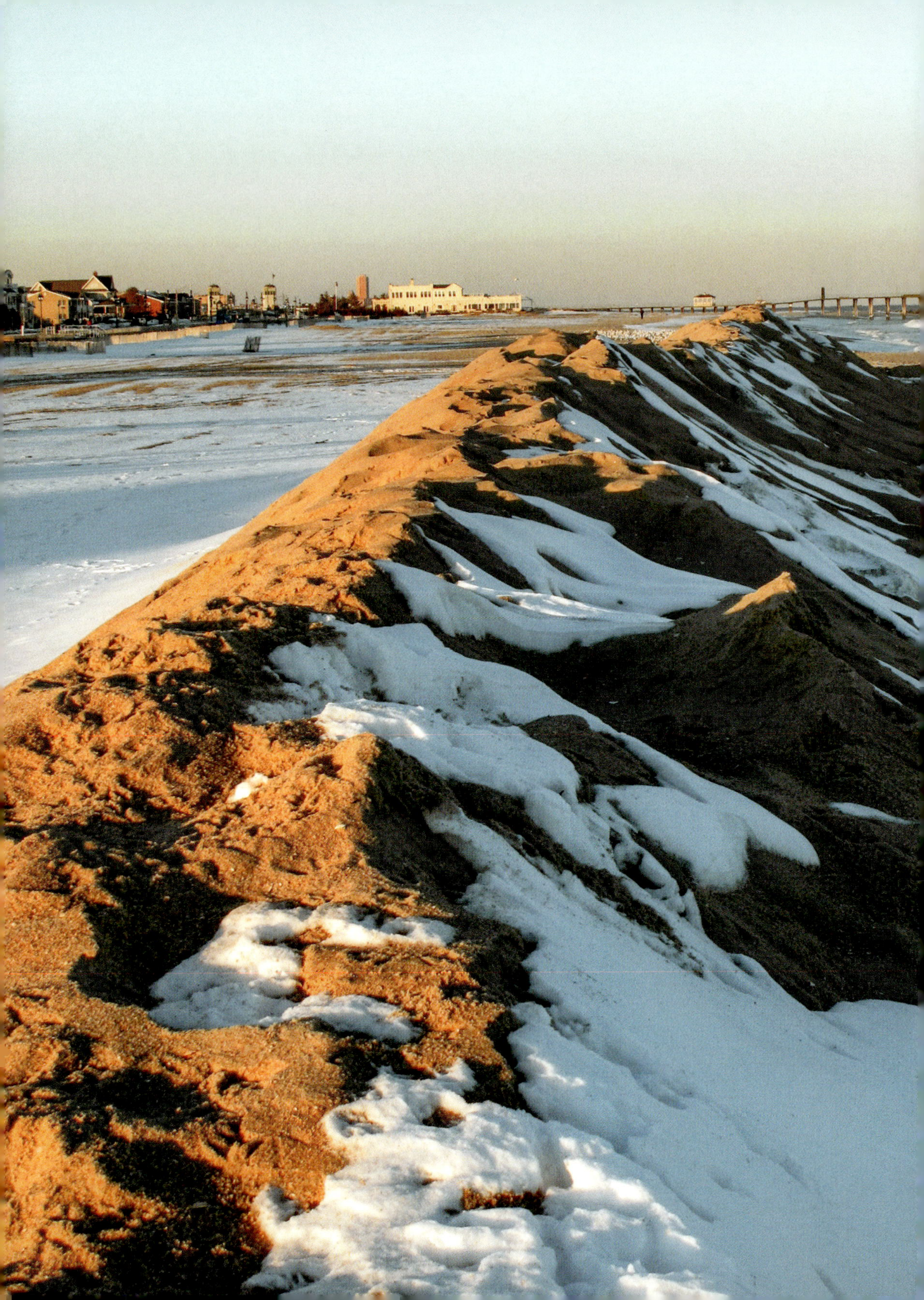

CHAPTER 4

SOFT ENGINEERING

Engineers don't always build structures to stop floods. Sometimes they use the land's natural defenses. This method is known as soft engineering. It is usually less costly than hard engineering. It is also better for the environment. This method can help improve water quality.

Engineers might dump extra sand on beaches to protect coastal areas from storms and flooding.

It also helps rivers and coasts adapt to **climate change**.

Healthy river **ecosystems** are less likely to flood. So, engineers work to protect these areas. For example, engineers may plant new vegetation near riverbanks. Trees and smaller plants help slow down floodwater. Their roots help hold the riverbank together. They keep the riverbank from **eroding**. A stable riverbank means the river will be able to hold more water. It will keep the land from flooding.

Engineers might use catchments to reduce the flow of floodwater. Catchments are areas where water

Healthy vegetation along rivers can help prevent flooding.

naturally collects. They include ponds and ditches. These areas store some of the floodwater. They also help break up the fast flow of a river.

Engineers also use soft engineering along coasts. They create living shorelines to make the shores stable.

These shorelines include natural materials such as plants, rocks, and sand. They provide a healthy home for wildlife. And they slow down waves during storms. In this way, the shorelines reduce erosion. The coasts are less likely to flood.

THE SAND ENGINE

For many years, people in the Netherlands relied on hard engineering for flood control. But in the 1990s, major flooding occurred. Engineers decided to try soft engineering methods. They built the Sand Engine. This hook-shaped area stretches along the coastline. Over time, waves and tides slowly move the sand onto nearby beaches. The sand prevents beach erosion and helps keep the area from flooding.

Catchments can store floodwater to keep other areas from flooding.

Soft engineering has some disadvantages. These methods need to be kept up often. Also, it can take a long time to grow trees and make new habitats. Sometimes, an area will flood again before engineers can finish their work.

CASE STUDY

HESKETH OUT MARSH

Salt marshes are areas along the seacoast where water covers the soil. In 1980, people built a seawall around the Hesketh Out Marsh in England. They turned the marsh into farmland. The wall protected the land against flooding. But sea levels kept rising.

The Royal Society for the Protection of Birds wanted to create more natural flood control. It bought the land. It decided to turn the land into a salt marsh again.

Engineers designed the new salt marsh. They decided where the water should come in. They developed creeks, banks, and lagoons. Engineers also thought about how the area would change over time.

Salt marshes are home to herons and other birds.

    Engineers completed the project in 2017. The new salt marsh acts as a barrier. It helps slow the tides and reduce the risk of flooding. It also provides new habitats for many birds and sea creatures.

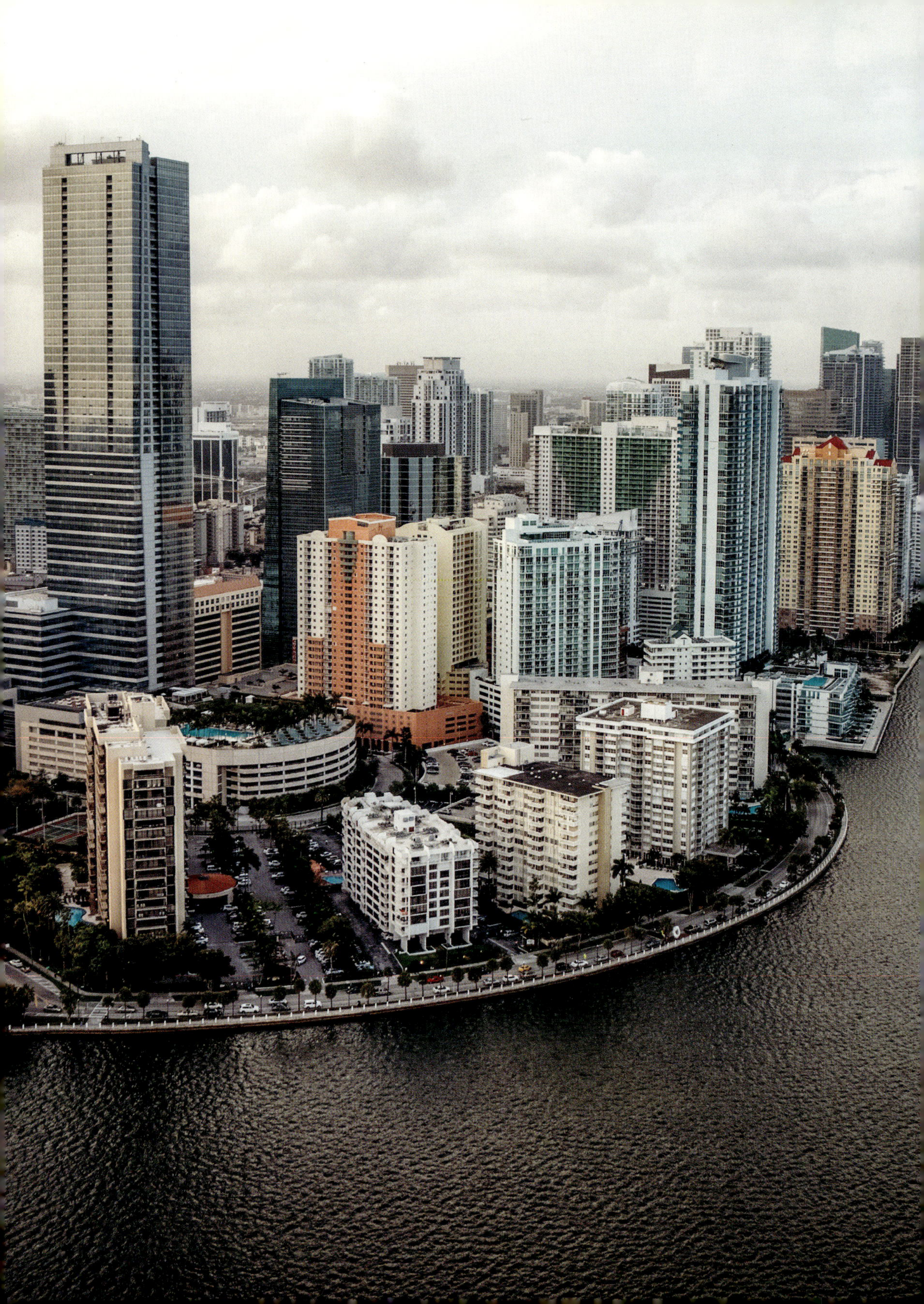

CHAPTER 5

PREPARING FOR FUTURE FLOODS

People can avoid flood damage by not living in areas that are likely to flood. These areas include flat land near rivers or streams. Areas near the coasts can flood easily, too. But many people want to live and work in these areas anyway. In fact, more and more people are moving into these areas.

Hundreds of millions of people around the world live in coastal cities.

So, engineers are figuring out ways to respond to floods more quickly. For example, one research team is using **satellites**. When a satellite passes over a flooded area, it uses **radar** to create a map. Engineers study the map. They see where the worst flooding is happening. They share this information with rescue teams. Emergency responders know where to go to help people in need.

Many engineers are combining hard and soft engineering. They see how nature can help reduce the effects of flooding. If they build structures, they do it with the environment in mind. In the future, engineers will continue to help communities adapt to and recover from floods. They will help keep people and property safe.

FLOODPLAIN GROWTH

A 100-year floodplain is a low-lying area that has a 1 percent chance of flooding in a given year. The Federal Emergency Management Agency studied 100-year floodplains. It predicted how population growth and climate change would impact flooding in these areas. This bar graph shows how 100-year floodplains are predicted to grow across the United States by the year 2100.

45%

Floodplains in river areas

55%

Floodplains in coastal areas

FOCUS ON
ENGINEERING FOR FLOODS

Write your answers on a separate piece of paper.

1. Write a letter to a friend describing the causes of floods.

2. Do you think people should be allowed to build new homes in areas at risk of flooding? Why or why not?

3. Which of the following is an example of soft engineering?
   - **A.** catchment
   - **B.** levee
   - **C.** drainpipe

4. How might erosion contribute to flooding?
   - **A.** Erosion helps the soil soak up water.
   - **B.** Erosion makes rivers flow faster.
   - **C.** Erosion removes the land that would act as a barrier.

Answer key on page 32.

GLOSSARY

channels
Stretches of water that connect two larger bodies of water.

climate change
A long-term change in Earth's temperature, air pressure, or wind patterns resulting from human activity or natural causes.

ecosystems
Communities of living things and how they interact with their surrounding environments.

eroding
The act of wearing away a surface.

levees
Walls built from earth materials to stop floodwaters.

radar
An instrument that locates things by bouncing radio waves off them.

satellites
Objects or vehicles that orbit a planet or moon, often to collect information.

vegetation
All of the plants found in a certain area.

TO LEARN MORE

BOOKS

Elkins, Elizabeth. *Investigating Floods*. North Mankato, MN: Capstone Press, 2017.

Felix, Rebecca. *Hurricane Harvey: Disaster in Texas and Beyond*. Minneapolis: Millbrook Press, 2018.

Thomas, Keltie. *Rising Seas: Flooding, Climate Change and Our New World*. New York: Firefly Books, 2018.

NOTE TO EDUCATORS

Visit **www.focusreaders.com** to find lesson plans, activities, links, and other resources related to this title.

INDEX

catchments, 20
channels, 14, 16
climate change, 20, 29

dams, 5, 10, 15
dredging, 16–17

ecosystems, 20

flash floods, 10
floodgates, 14
floodwalls, 7, 13

Great Flood, 6–7

hard engineering, 13–14, 16, 19, 22, 28

levees, 5–7, 13
living shorelines, 21–22

pavement, 14

rain, 5–6, 9–10, 14, 16
runoff, 9

salt marshes, 24–25
Sand Engine, 22
satellites, 28
soft engineering, 19, 21–23, 28
soil, 9–10, 13–14, 24

vegetation, 10, 20

Answer Key: 1. Answers will vary; 2. Answers will vary; 3. A; 4. C